THE MAKING OF **YNYSLAS**

TALES FROM AN AREA THE SIZE OF WALES — 25,000 YEARS AGO TO THE PRESENT DAY

JOHN S MASON

Word after Word

Published by Word After Word Press, 2019
FE20191

Word After Word Press
www.wordafterwordpress.com

ISBN 9781916165502

A CIP catalogue record for this book is available from the British Library.

Design and layout by Sara Holloway Graphic Design - www.sarahollowway.com
Printed and bound by Instantprint

Thanks to Prof. Chris Michael of Liverpool University, for the digitised map on page 6.
Further such images may be examined at:
https://www.liverpool.ac.uk/~cmi/lewismorris/lewis.html

In memory of

Shena Mason

(1938-2014)

CONTENTS

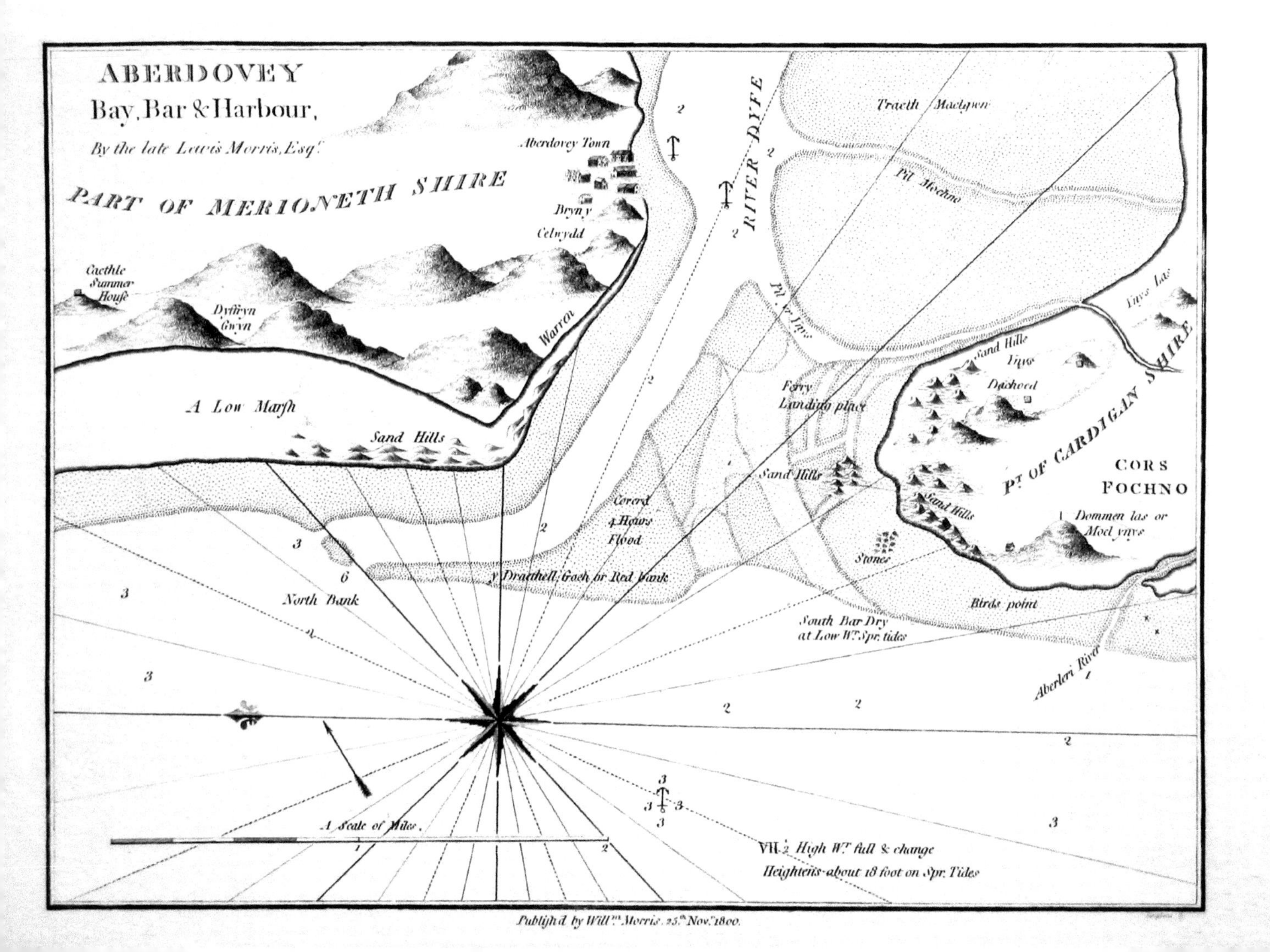

Publish'd by Willm. Morris 25th Novr. 1800.

INTRODUCTION

Ynyslas, translating from Welsh as "Blue Island", is situated near Borth, in western Mid-Wales, between the confluence of the major river, Afon Dyfi, with its smaller tributary, Afon Leri, and the open shore of Cardigan Bay.

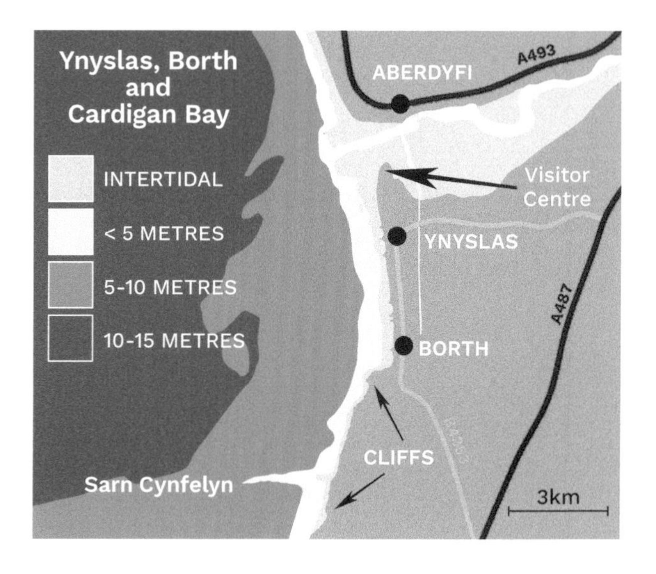

Within this small patch of Wales' geography, there exists a remarkable diversity of natural ecosystems, from intertidal sands to saltmarsh, from shingle storm beach to dry sand dunes and the damp slacks that lie between them. Each of these different habitats supports its own specialised flora and fauna, adapted to conditions that would defeat less robust species.

Along the open shore, there is the famous Submerged Forest, visible at low tides and a relic of the distant past, when sea levels were lower. Indeed, the Forest is but a fragment of a once extensive wooded plain, now occupied by Cardigan Bay and drowned as the world's great ice caps melted following the end of the last ice age.

This is an account of that drowning — when it happened, how quickly, and how Ynyslas itself came into being and is maintained by natural processes to this day. On the geological scale of time, the place is but a youngster.

Much of Ynyslas is a National Nature Reserve, covering the sand dunes, some of the storm beach and most of the Dyfi Estuary. The place is immensely popular, welcoming over 250,000 visitors every year — some coming for the wildlife, others for gentle recreation and still others to lounge in the sunshine in deckchairs by their car. There's something here for everyone, it seems, apart from swimmers, for the rip tides around the Estuary and its mouth are notoriously ferocious.

Natural Resources Wales manages the Reserve, whose staff have the interesting task of balancing the need to protect habitat, flora and fauna, with the varied expectations of the visiting public. There is a visitor centre, close to the edge of the sand dunes, where people can find out more about the Reserve, its wildlife and history. Activities and events, organised by the staff throughout the summer months, include guided walks. The centre is open from spring to mid-autumn, and is well worth a visit.

Looking up the Dyfi Estuary from Ynyslas dunes.

SETTING THE SCENE: Our current Icehouse climate

Ynyslas is a direct product of dramatic changes in Earth's climate that have taken place over the past 25,000 years. These changes drove the formation of the landscape as we see it today: a long, storm-beach-fronted shingle spit, topped with sand dunes, and the sandy estuary and muddy saltmarshes beyond. But these familiar features were not always there. How do such changes come about in the first place? Here is a quick climatological crash-course.

Through geological time, Planet Earth has variably experienced two different climate states, known as Hothouse and Icehouse. In a Hothouse climate, Polar ice caps are absent, whereas in an Icehouse climate, they are always present. Not only that, but in an Icehouse climate, there are periodic, cyclic spells, lasting for tens of thousands of years, in which conditions become even colder. Such spells are known as glaciations, or ice ages, during which the ice sheets expand from both Poles towards the Equator. Glaciations, alternating with milder interglacials, formed the cyclic climate pattern throughout the Quaternary, a division of geological time that started 2.58 million years ago. We live in the latest interglacial.

The cyclic patterns were driven by regular variations in Earth's orbit around the Sun, known as the Milankovitch Cycles, each lasting tens of thousands of years and affecting how much sunshine and therefore how much energy the planet receives. Named after the Serbian physicist, Milutin Milankovitch, who proposed their existence about 100 years ago, such periodic variations have likely been a feature of planetary climate for countless millions of years. They would likewise have made a Hothouse climate swing between pleasant barbecue weather and unbearable, sweltering heat over similar time-spans.

What is intriguing is that going even further back into the depths of geological time, we only see sporadic evidence for periods of glaciation. In fact, the Icehouse climate state is not that common and special circumstances must prevail to force a transition away from the dominant Hothouse.

What drives Earth's long-term climate?

There are two main drivers of long-term climate on Earth.

One is the energy that comes in from the Sun. Fortunately for us, our Sun is a stable star that steadily brightens during its 10 billion year lifetime (known to astronomers as its "main sequence"), as it works through its supply of hydrogen fuel. The current rate of increase in solar luminosity, as it's known, is about one percent per 100 million years – not noticeable in human life-spans or even from generation to generation. Apart from the relatively minor variations caused by things like sunspot cycles, the brightness doesn't jump around everywhere. Instead, with stars of this type it's a slow but steady climb. Were it otherwise, Earth's climate state would have been perpetually chaotic, which would have completely messed with the progressive evolution of advanced life.

The second driver involves the fate of all that incoming solar energy once it has arrived here on Earth – how much of its energy is retained and how that energy is moved around on the planet. It was as long ago as the 1820s that it was first guessed that there had to be something like an insulating blanket, to trap heat within the atmosphere. Without such insulation, a planet of this size and at this distance from the Sun, ought to be frozen solid. Incidentally, the distance from the Earth to the Sun had already been calculated as early as 1771, albeit not quite correctly. The figure obtained was 153 million kilometres, whereas the correct value, measured with modern technology, is 149.6 million kilometres. Not a bad attempt for the late 18th Century.

Later in the 1800s, the vital roles of greenhouse gases such as carbon dioxide (CO_2) were discovered and investigated. By the late 1890s, the Swedish scientist Svante Arrhenius had calculated the warming effect that would occur if atmospheric CO_2 levels, then at around 300 parts per million (ppm), were doubled. His result was remarkably close to modern thinking. But in the 1890s, emissions of CO_2 were relatively low and Arrhenius could not foresee a point when that doubling would occur. Since then, industrialisation has expanded so rapidly that CO_2 levels have now passed 415 ppm and are increasing at more than 20 ppm per decade. These levels are now higher than at any time in the past three million years.

Chaotic sediments left behind by melting glacial ice, on the Cardigan Bay coast at Tonfanau, north of Tywyn.

So, on a geological time-scale, we know that the Sun is warming up slowly and its year-to-year fluctuations have small effects on our climate. But much longer cycles involving Earth's orbit have profound impacts over tens of thousands of years. We also know that large changes in atmospheric greenhouse gas concentrations have major climate effects. How do such changes fit in with what actually happened on Earth, in its past?

Icehouse climates always leave behind them geological fingerprints that are unmistakeable to the trained eye, such as the rubbly, unsorted mish-mash of rocky debris left behind when a glacier melts. Compared to the neatly-bedded, well-sorted water-lain sedimentary rocks, such as sandstones and conglomerates, glacial deposits look more like they have been fly-tipped. Rocks composed of such distinctive material, along with a host of other evidence, mark the comings and goings of periods of deep cold, back through geological time. We can therefore pinpoint such freeze-ups, going back hundreds of millions of years.

Let's take a look down the time-line.

Figuring out Deep Time

First things first. With geological time, we are talking about millions of years ago. Set against familiar things like our day-to-day lives, such numbers are difficult to visualise. A similar problem arises when trying to imagine light-years, as units of distance in space. In each case, we get taken clean out of our frames of reference.

What we need is context. A good work-around is to plot geological time, from the formation of Planet Earth through to the present day, against something much more familiar, like the human 12-month calendar.

If we start our dual time-line with Earth coming into being 4,500 million years ago, a millisecond into the morning of New Year's day, we can then go right around the calendar until the bells are about to ring on New Year's Eve, which, geologically-speaking, is "now".

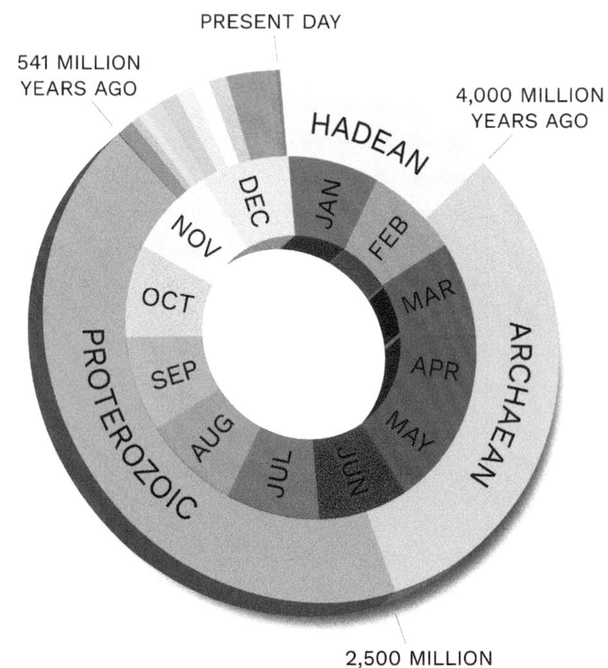

Looking at geological time in this way, we can see the progression of the billions of years that make up Earth's long past and relate them to specific parts of our year – days, months, seasons and so on. As will be revealed below in more detail, Earth's atmosphere first became oxygenated just after mid-June. The land was first colonised by green plants in late November and the dinosaurs started to roam the planet in mid-December, but then promptly died out on the evening of Boxing Day, just as Jurassic Park was about to start on the TV.

For the purposes of our story, let's begin in mid-June, in other words some 2,500 million years ago, on an inhospitable Earth where, if you were to land on its surface, you wouldn't last more than a matter of seconds without breathing equipment. The atmosphere of Earth back then, at the start of the Proterozoic Eon, was an unappealing mixture of nitrogen, carbon dioxide, water vapour, hydrogen, carbonyl sulphide, ammonia, methane and other gases.

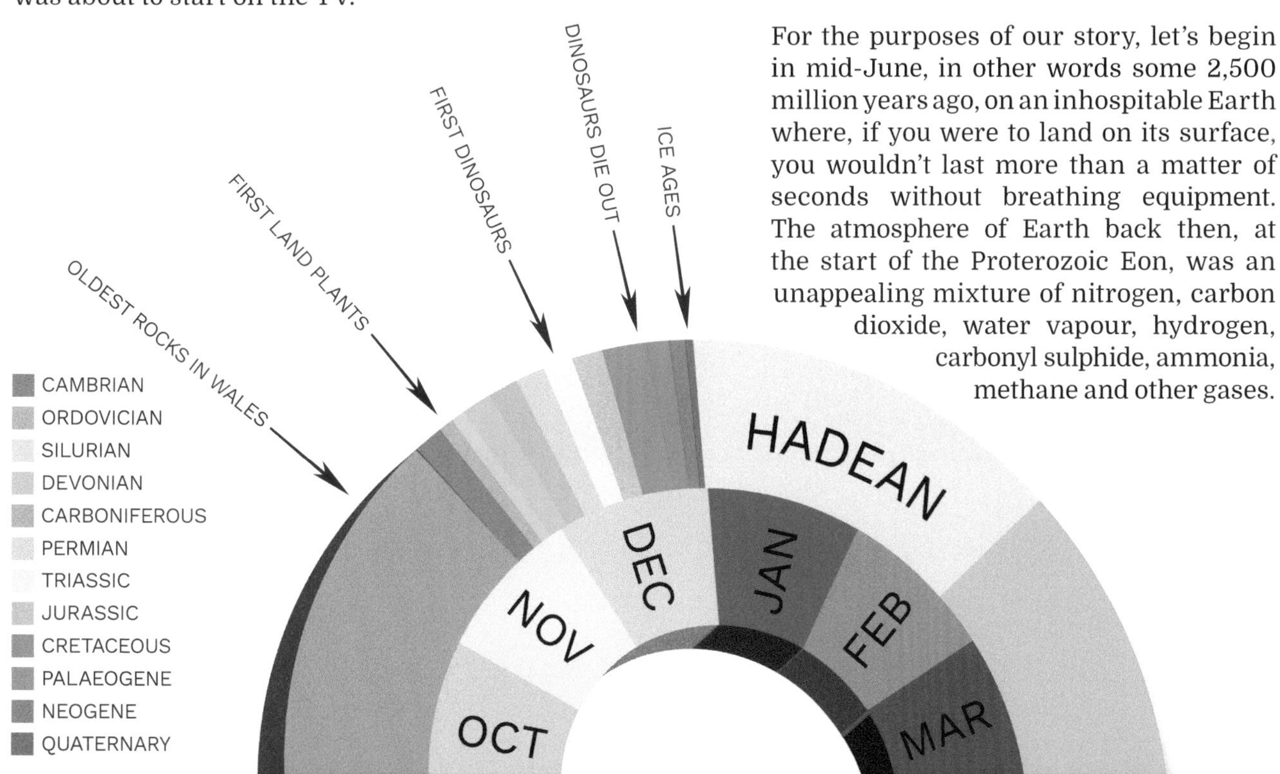

A fundamental change on Earth: the Great Oxygenation Event

The Great Oxygenation Event, as it is known, began some 2,450 million years ago. Of all the big changes that have happened through geological time, few have had such profound importance as this one. Occurring progressively over some 250 million years, this was an atmospheric point of no return. In its wake, the rate of supply of oxygen to Earth's atmosphere outstripped its rate of consumption, so that Earth now had an oxygenated atmosphere. So how did this remarkable event happen?

Atmospheric gases have both "sources", from which they are supplied, and "sinks", into which they become locked-up. The locking-up of oxygen occurs due to chemical reactions. For example, at Earth's surface, metallic iron readily reacts with atmospheric oxygen in the presence of moisture. The product of that chemical reaction, familiar to all, is rust, which is a form of iron oxide, a fairly stable substance that will last for a long time. Rust is therefore a good example of an oxygen sink.

At the beginning of the Proterozoic, the scarcity of free oxygen in Earth's atmosphere would have prevented that rust-forming reaction from taking place on land. Yet iron is a very common element on Earth. Before the Proterozoic, iron used to be present in enormous amounts in the oceans, dissolved in the seawater. But at some point back then, a rather special life-form arrived on the scene: green slime, officially known as "oxygenic photosynthesising cyanobacteria". In oxygenic photosynthesis, an organism absorbs carbon dioxide and water and, with the help of sunlight, combines them in a series of chemical reactions to make sugar. That's how they feed. The other product of these reactions, oxygen, is not required and is released back into the environment.

In that ancient world, those photosynthetic bugs found the warm, shallow and sunlit seas around the continents much to their liking. They were probably a bit like that fast-growing green, filamentous blanket weed that can clog up garden ponds in a warm summer. The cyanobacteria continuously released tiny bubbles of oxygen into the iron-rich seas in which they thrived. In turn, the oxygen reacted with the dissolved iron in the water, making insoluble iron oxides, which would then sink, in countless tiny particles, down to the seafloor. As the years, decades and centuries passed, great deposits of iron oxide – known to geologists as "banded iron formation" – slowly accumulated in this manner. Banded iron formation was, therefore, an enormous example of a long-term oxygen-sink, on an unimaginable scale when compared to a rusty exhaust pipe.

Around 2,450 million years ago, something of extreme importance happened. So much of the dissolved iron had by then been gobbled up, in the above manner, that the tiny bubbles of oxygen now had a much better chance to rise up through the water and enter the atmosphere. As dissolved iron levels fell and fell, so the deposition of banded iron formation dwindled and the atmosphere became increasingly oxygenated, as that photosynthetic slime continued to thrive in the waters below. This was to be a permanent change on Earth, and with other consequences, too. Prior to the Great Oxygenation Event, methane, an extremely potent greenhouse gas, was a significant component of the atmosphere. The arrival of abundant oxygen put pay to that, since it readily oxidises methane into carbon dioxide and water vapour – both greenhouse gases too, but with a lot less clout. With a significant reduction in the overall greenhouse effect, temperatures plummeted, marking the onset of the Huronian glaciation, considered by geologists to be the first global Icehouse climate event to have affected Earth.

Recovery from the Huronian deep-freeze could no longer be helped by methane. Instead, it was the constant supply of carbon dioxide from volcanic activity that would have eventually tipped the balance in favour of warmth again. Today, according to the U.S. Geological Survey, volcanoes emit up to 300 million metric tonnes of carbon dioxide every year. This might not sound like much, compared to human fossil fuel-related emissions of more than 30,000 million tonnes a year. But – and it's a big but – if the mechanism for adding CO_2 to the atmosphere stays switched on, and the mechanism for removing the gas gets switched off, its atmospheric concentration will build up rapidly through time.

Let's take a quick look at how carbon dioxide is removed from the atmosphere. Today, photosynthetic plants use a lot of the gas to make food, and they are responsible for the regular wiggles on modern plots of CO_2 concentration — the downticks represent the growing season in the Northern Hemisphere, where the majority of Earth's landmasses happen to be. Some CO_2 dissolves directly into the oceans, at the water-air interface. But on geological time-scales, the most important process is the chemical weathering of rocks.

Mauna Loa observations, 1999 to mid-2019. Data: NOAA

Atmospheric carbon dioxide concentrations for the past twenty years, showing the rhythmic wobble due to the Northern Hemisphere's growing season.

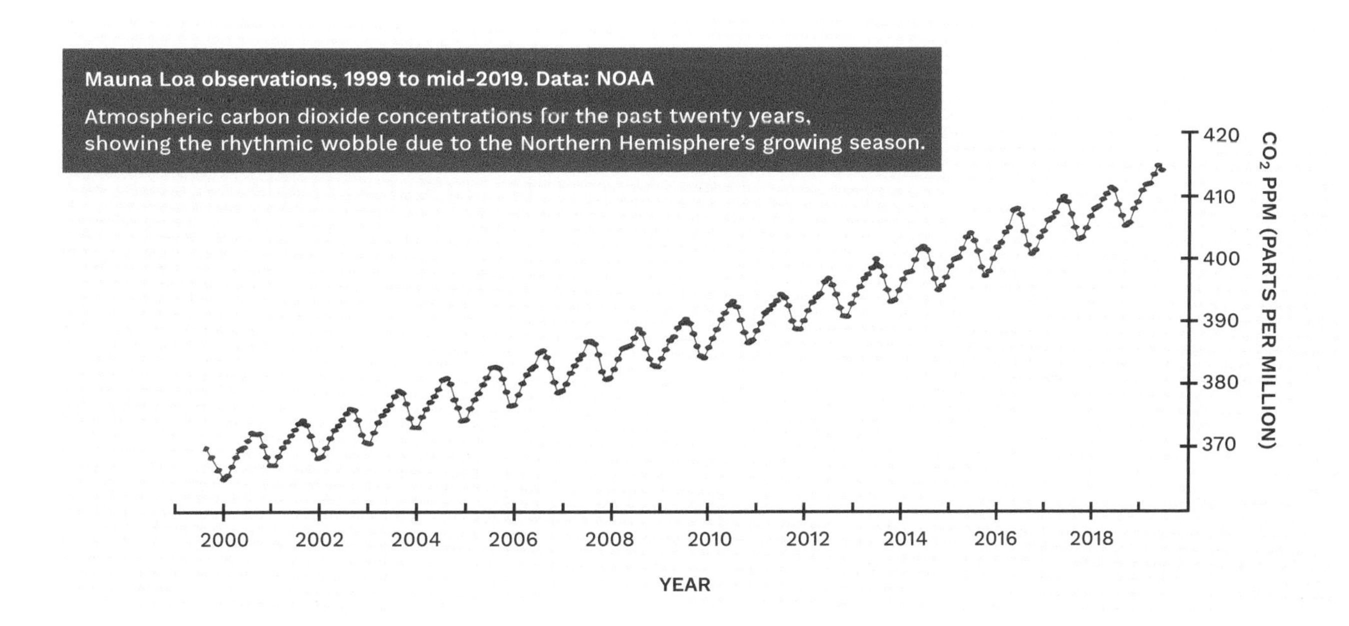

Weathering, a vital part of what we call the Slow Carbon Cycle, involves the slow breakdown of solid rock, due to the chemical reactions that take place between the various minerals making up a rock and weathering agents. The most important of the weathering agents is CO_2 dissolved in rainwater — or carbonic acid. Different minerals vary in their stability when exposed to this weak acid, so that some will dissolve quickly, others very slowly. Given enough time, though, weathering can convert an outcrop of rock into a loose, gritty, clay-rich residue of insoluble minerals. Water-soluble weathering products, such as sodium, calcium and bicarbonate, go into solution in the form of charged particles called ions. The solutions are flushed away by rainwater, into the rivers and down to the oceans, a process that has been going on throughout geological time. It's why the sea is salty.

Weathering of rocks — an action shot. This hand-sized sample of grey basalt, from North Wales, has a paler buff to rusty-looking rind around its outside surface, some five millimetres in depth. Here, the surface of the rock has been exposed to the weather and the paler colour of the rind is due to weathering. In the rind, the darker iron, calcium and magnesium-rich minerals have been slowly attacked by carbonic acid (CO_2 dissolved in rainwater) and dissolved away, leaving behind a pale residue of relatively insoluble minerals and iron oxides. Because this sample is from a glacially-eroded outcrop and we know the great ice sheets of the last glaciation had retreated from Wales by 19,000 years ago, we can say that this pale rind is the result of about 19,000 years of weathering. It is a slow process. But on geological time-scales, millions of years of weathering would convert the entire sample to the same, pale bleached-looking appearance. Now imagine the same thing happening to entire mountain ranges. →

The weathering process, a vital part of Earth's slow carbon cycle, uses up a lot of CO_2 and enriches the oceans with sodium, potassium, calcium, magnesium and bicarbonate ions, these being the water-soluble products of the chemical reactions that take place.

Many kinds of sea-dwelling plants and animals build their shells or reinforce their tissues by extracting dissolved substances from the water. For example, sea shells, like the ones that can be found along the beach at Ynyslas, consist of calcium carbonate ($CaCO_3$). To make this substance, shellfish extract calcium and bicarbonate ions from the seawater and combine them together. In some circumstances, calcium carbonate can also form directly, as a chemical precipitate, without the need for a living middle-man. Either way, calcium carbonate-rich sediments accumulate on the seabed over decades and centuries, eventually hardening into the solid rock we call limestone. In that way, some of the carbon that was originally in atmospheric CO_2 becomes locked away, stored as calcium carbonate in limestone, for many millions of years.

In the Huronian, though, there was a problem. As is the case with many kinds of chemical reactions, the warmer the conditions, the faster the weathering. The reverse of that is also true, so that when it comes to weathering, a deep-frozen planet is pretty useless. With weathering pretty much switched off, but business-as-usual for the volcanoes, there could only be one outcome: the amount of atmospheric CO_2 built up and up and up, eventually restoring a strong greenhouse effect, and the Hothouse climate then returned.

← Storing ancient carbon: this 180 million year old Jurassic ammonite shell and the limestone surrounding it are both made of calcium carbonate.

Cryogenian - in the deep-freeze

Earth basked in warmth for hundreds of millions of years following the end of the Huronian Icehouse. The chain of events that led to the onset of the next cooling, leading to what we call the Snowball Earth glaciations, did not begin until around a billion years ago — or October, on our 12-month calendar.

What was special about that time? Various hypotheses have been offered over the years. We do know that a great supercontinent was situated over Earth's equatorial regions. Geologists like to have names for such things and they call this one Rodinia. Continents, like America or Africa, may seem too big for anything to happen to them, but this is not so over geological time-spans. Rifting can occur, when one of Earth's tectonic plates comes under prolonged and major tension, so that it splits into two or more pieces, separated by deep fissures that go all the way down to the bottom of Earth's crust. When a continent rifts, magma then flows up the fissures, to erupt as molten lava at the surface. Rodinia's break-up, beginning some 850 million years ago, was accompanied by gargantuan outpourings of lava, which upon cooling and solidifying formed the igneous rock, basalt.

Basalt is a dark-coloured, dense rock, consisting of minerals that contain lots of iron, magnesium, calcium and sodium, all especially susceptible to chemical weathering. Making vast amounts of basalt available to weathering agents, in this warm, equatorial climate, would have created an ideal scenario for the process. As a consequence, there would have been a correspondingly major fall in atmospheric carbon dioxide levels, and a weakening of the greenhouse effect, over the following tens of millions of years. As ice sheets began to grow, their bright surfaces increasingly reflected incoming sunlight back out into space, reducing the amount of energy received by Earth and letting temperatures fall still further.

The resulting Snowball Earth glaciations occurred 720 to 635 million years ago, during a period of geological time that commemorates them in its name, the Cryogenian. Glaciations were extreme, almost global in extent, but yet again, because of the extreme cold, the processes that draw down atmospheric CO_2 slowed right down, whereas volcanic activity continued unabated. Each of these glaciations seems to have ended very quickly. With massive-scale melting of the ice, the rising seas quickly flooded in over the land in many parts of the world. In these seas, limestones known as “cap carbonates”, were deposited directly on top of the distinctive glacial rubble. That glacial rubble-cap carbonate sequence, found in many places worldwide, is the distinctive geological hallmark of the Cryogenian.

Transition from a glacially-deposited, boulder-rich sedimentary rock (tillite, below), to well-layered pinkish and yellowish cap carbonates (above), marking the typical, rapid end of a Snowball Earth episode, 635 million years ago in the late Proterozoic. →

This rock outcrop is in the Kaokoveld, Namibia. Paul Hoffman, a key "Snowball Earth" researcher is pointing to the tillite-cap boundary. Photo: M.J. Hambrey

Climate change and mass extinction: the late Ordovician disaster

After the Cryogenian, warmer conditions prevailed for almost 200 million years. By then, life had evolved to a far more advanced state, well beyond the mostly simple, single-celled forms of the Proterozoic. An evolutionary explosion had taken place in the seas, during the Cambrian Period, (541 to 485 million years ago), leading to the appearance of all sorts of complex, multicellular organisms. As a consequence of this kick-start, the marine environment of the subsequent Ordovician Period (485 to 443 million years ago) was teeming with life, some of which would have looked similar to modern forms. A particularly welcoming habitat was to be found in the warm, shallow, well-lit nutrient-rich waters fringing Earth's continents. The fossil record from the late Ordovician shows that there were abundant shellfish, dwelling in burrows in the sand or clustered together clinging to underwater rocks, around which primitive fish darted and trilobites crawled and foraged. Coral reefs flourished and were home to a tremendously rich biodiversity, yet at the same time they were incredibly prone to sudden environmental change. The rapid onset of an intense period of continental-scale glaciation, 445 to 443 million years ago — or near the end of November on our dual time-line — proved to be disastrous.

Earth's return to Icehouse conditions is considered to have been triggered by atmospheric CO_2 depletion, caused firstly by a runaway weathering episode and secondly by the colonisation of the land surface by primitive photosynthetic plants. Accompanied by a rapid and major fall in global sea levels, massive marine habitat-loss clearly occurred. Ocean water chemistry was also affected, with widespread evidence for the development of low oxygen levels and the presence of toxic hydrogen sulphide. Due to such changes, the mass-extinction was the second biggest of the past 500 million years. The exact kill-mechanisms are still debated, but what is certain is that this event was little short of catastrophic. Just as an example, it took as long as six million years for coral reefs to reappear, showing that, while some corals did manage to survive, they were nevertheless decimated.

If you look inland from Ynyslas, up the Dyfi Valley, the hills on either side are made up of sandstones, siltstones and mudstones that date from the late Ordovician and early Silurian Periods. Now hard, grey rocks, tilted and folded by earth movements, back then at the Ordovician-Silurian boundary, they were soft sediments, deposited layer-upon-layer on an ancient seabed. Features within these rocks, such as sudden changes from fine to coarse sediment, reflect the abrupt physical changes that took place during that dreadful time. The earliest Silurian rocks include dark-coloured shales, full of iron pyrites, a mineral only stable in seafloor sediments when the adjacent water's oxygen levels are very low, thereby reflecting changes for the worse in water chemistry.

Sandstones, siltstones and mudstones of late Ordovician age in the Cambrian Mountains of Mid-Wales: now solid rock, but 445 million years ago these were soft sediments.

Cooling and coal swamps

Another much longer cold period, lasting several tens of millions of years, began in early December — some 360 million years ago, in the Carboniferous Period. This one came hot on the heels of the greening of the lands of Planet Earth.

The first land plants known from Wales grew in the late Silurian Period, some 420 million years ago. These are their remains. Field of view: just five millimetres wide. They were tiny things. Image: Diane Edwards

The colonisation of the land by photosynthetic plant life had slowly begun in the Ordovician, but those pioneers were lowly things, a bit like the mosses and liverworts we see today, covering damp rock-faces. By the Carboniferous, though, tens of millions of years of plant evolution and diversification — and, vitally, soil formation — had left large parts of Earth's landmasses covered in lush forests.

Two things happened as a consequence of this great explosion in green plant life. Firstly, Earth's atmosphere became more oxygenated than at any other time in our planet's history, with around 35% oxygen, as opposed to today's 21%. Secondly, all those plants gobbled up a lot of carbon dioxide, so much so that the atmosphere's heat-retaining properties eventually became diminished and Earth's climate again cooled down: for a time, the Icehouse was back.

Wales was not in a suitable geographical position to be covered by the Carboniferous ice sheets. The layout of Earth was very different back then, thanks to plate tectonics and continental drift. An ancient continent, Avalonia, of which Wales was a part, had been drifting northwards through the Southern Hemisphere for the previous 200 million years. Continental drift is a slow process – at its fastest, you are looking at a similar rate of movement to that at which human fingernails grow. But by the late Carboniferous, Avalonia had arrived somewhere near the Equator, a warm part of Earth even in an Icehouse climate. Vegetation thrived so luxuriantly that great thicknesses of peat accumulated in vast tropical wetland forests – the coal swamps. The thick layers of peat, progressively buried ever more deeply beneath further major influxes of waterborne sediment, would go on in due course to become the coal seams of South Wales and elsewhere: a vast store of fossil carbon, trapped in time, only to be mined by people some 300 million years later.

Since the Carboniferous Period, the only major glaciations to have affected the mid-latitudes, such as Europe, have been those of the past 2.58 million years. For much of the rest of that time, the Hothouse climate state has prevailed. The descent into this ongoing Icehouse climate started in earnest some 33 million years ago, when the first ice sheets appeared on Antarctica. Its cause is becoming better understood and is thought by many scientists to be related to major mountain building, caused by numerous continental collisions over the past few tens of millions of years. During these events, great ranges such as the Alps and Himalayas were created. Such geological violence would have led to an enhanced weathering-regime, drawing down more CO_2 from the atmosphere.

By the late Carboniferous Period, 310 million years ago, parts of Wales were cloaked in dense forests made up of tree ferns and other species. This fossilised fragment of a fern frond is ten centimetres in length.

That quick tour through deep time shows the importance of large shifts in Earth's Carbon Cycle in driving sometimes wild fluctuations in global temperature. Such fluctuations are far more prolonged than our ever-changing weather or even seasons: they determine the habitability, over millions of years, of large parts of the planet's landmasses and oceans.

So, now we have the context: we are living within an Icehouse climate, one of a small number of such climatic states dotted through geological time. Our time is within an interglacial: the intense cold of the last Glacial peaked some 25,000 years ago. Afterwards, a sequence of natural processes brought Ynyslas into being.

We can now begin the story proper of how Ynyslas was made, although 25,000 years still seems like a long time ago. Where, then, does it appear on our dual time-line?

At one and a half minutes before midnight, on December 31st.

THE MAKING OF **YNYSLAS**

Part one: 25,000 to 11,784 years ago - Meltdown

We start the tale of Ynyslas 25,000 years ago because it is a product of all of the physical processes that have happened since. In fact, without such processes, there would be no Ynyslas. Climate change, melting ice sheets, sea level rise, collapsing sea cliffs, violent storms: all have had vital roles. But 25,000 years ago also marked an important milestone in the evolution of the Welsh landscape, during the time known to Earth scientists as the Last Glacial Maximum.

Imagine, for a moment, being in Wales, back then:

It is a bitterly cold but clear day. Looking out from Yr Wyddfa (Snowdon) in every direction, I see ice. There is ice everywhere. Great crevasses yawn open where the upper glaciers pull away from the steep, frost-shattered black mountainside. To the south, the ice cap stretches into the distance, punctuated by an isolated summit here and another there, jagged dark teeth stabbing up from the seemingly endless white plateau. To the west, glaciers head out across the great plain to merge with the vast, southwards-flowing ice sheet occupying the space where the Irish Sea should be.

Not the friendliest of places, then. Apart from the very far south, nobody could have lived in Wales at that precise point in time. Yet the deep-freeze was at last, after many thousands of years, coming to an end, and compared to the long cooling down to the Last Glacial Maximum, once started, the end came quickly.

Climate feedbacks: causes and consequences

We have seen how, once the planet is within an Icehouse climate state, the advances and retreats of the ice sheets, towards the Equator and back towards the Polar regions, are initiated by cyclic changes in Earth's orbit. This was also the case after the Last Glacial Maximum. But there's another important factor: feedbacks.

In the physical world, feedbacks are things that are forced to happen because of other physical changes. Feedbacks may then go on to cause those same changes to become amplified or reduced in their strength. As an example, snow and ice reflect a lot of incoming sunshine, which is why on a bright day on a ski-slope, you need tinted goggles to avoid getting dazzled or worse. Much of that reflected sunlight heads back out into space, where it is not a lot of use in warming up the planet. The more extensive the snow and ice cover, the more reflected sunlight and the more cooling — and the more extensive the snow and ice cover. That's a classic example of a feedback at work.

When a glaciation comes to an end, the opposite happens: the snow and ice cover retreats and the frozen soil beneath sees sunlight for the first time in many centuries. Soil is much darker than snow and instead of reflecting, it absorbs most of that incoming sunshine and warms up. As the soil warms up it thaws out, so that gases like carbon dioxide and methane, produced by soil bacteria before everything froze solid, are released back to the atmosphere. Imagine that happening not just to one patch of melted snow, but to hundreds of thousands of square kilometres of permafrost: that's a lot of gas being released. Because carbon dioxide and methane are greenhouse gases, the planet will then start to warm more quickly. We know this happened, because we can analyse samples of the atmosphere, trapped for thousands of years in glacial ice and accessed via ice-cores, drilled from the Greenland ice cap, Antarctica and elsewhere. We can therefore track how atmospheric composition changed, as the glaciations came and went over hundreds of thousands of years.

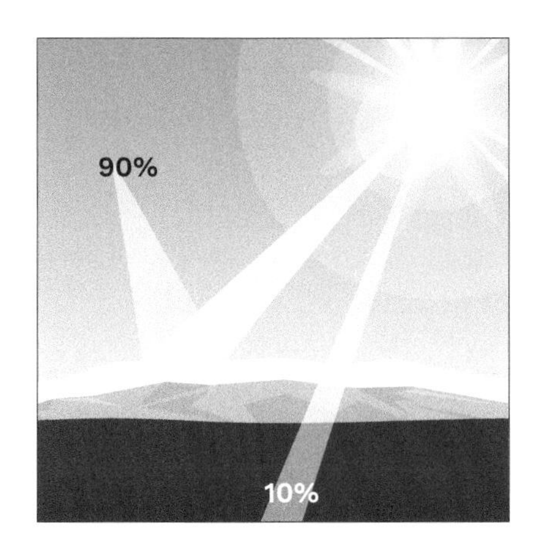

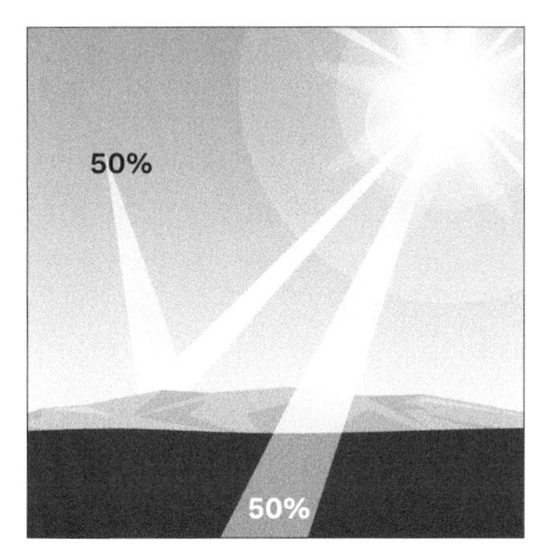

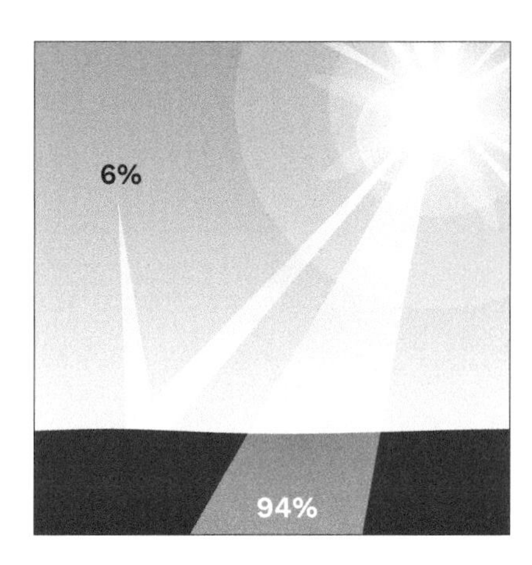

Another example of feedbacks. Freshly-fallen snow is highly reflective of incoming sunshine, so that most of the solar energy is simply bounced back towards space. But sea ice can absorb about half of the incoming energy, so if conditions become warmer, causing the snow to melt, there's more energy retained on Earth. If the sea ice melts too, then almost all of the incoming solar energy is absorbed by the much darker surface of the sea. By this feedback mechanism, an initial warming directly results in further warming.

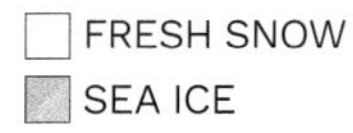

Amplifying feedbacks, once triggered, were the additional drivers that made the transitions from glacials to interglacials a rapid business, in geological terms. As the climate warmed, slowly at first then more quickly as feedbacks kicked-in, ice caps over the Welsh mountains shrank and the glaciers retreated. Glacial ice is one of the most powerful agents of erosion on Earth. Glaciers carved out the familiar Welsh landscape of today from solid rock. Then, when they melted away, the glaciers left behind them countless billions of tonnes of rock debris. There would have been piles and banks of rock fragments, ranging from boulders several metres in diameter all the way down to tiny particles, clogging the valleys and strewn out over the plain. Meltwater rivers would have thundered their way along the valleys, especially during the summer months when the thaw was at its most rapid. They would have dwarfed the familiar Welsh rivers of today. Out over the western plain they would have flowed, into lakes or straight out to the coast, transporting and depositing sands, gravels and, in more sluggish stretches, silts and muds. There would have been enough raw material for as many beaches as you could have wished for.

Sarn Cynfelyn or The Causeway, as it is often called, is a long ridge of glacially-dumped boulders and cobbles, that extends out some six nautical miles into Cardigan Bay. Its landward end dries out at low tide, but the rocky shoals formed by its underwater continuation can be seen here as a darker disturbance on the otherwise flat sea surface. The disturbance is caused by the tide being forced up over the ridge. There are several such features along the Cardigan Bay coast, all relics of a time when the area would have been unrecognisable. →

Sarn Cynfelyn, viewed here from the Wales Coast Path, between the villages of Borth and Clarach.

Let's have a quick stock-take of Wales, a few thousand years after the Last Glacial Maximum. Mountain ice caps had begun to dwindle and the Irish Sea ice sheet was in retreat. Retreat is perhaps an understatement: research indicates it was more like a rout. This huge, thick sheet of ice, reaching at its peak all the way down to the south-west of the Isles of Scilly, was to collapse so quickly that by about 19,000 years ago, its front is thought to have been situated to the north of the Isle of Man.

In the aftermath of the demise of the great ice sheet, the Irish Sea would have been a bleak, low-lying wasteland of mud, sand, gravel and boulders. The actual coast would have still been distant: it was well to the south-west, beyond the Celtic Deep off Pembrokeshire. Rivers, flowing out across these lowlands on their way to the distant sea, carved deep channels through the ice-deposited debris. In the Mawddach Estuary near Barmouth, such channels have been geologically investigated, through geophysics and boreholes. They were found to be between 24 and 43 metres deep. The Dyfi Estuary is likely to have similar, deeply buried features, since in 1981, a borehole was drilled near the northern end of Ynyslas dunes. It went down through 30 metres of soft, post-glacial marine sediments, without ever reaching bedrock.

It's hard to imagine now, if you stand on Barmouth Bridge, that 19,000 years ago, the river would have been flowing tens of metres beneath you. But this hints at the scale of the changes involved in the making of Ynyslas: they were on a monumental scale.

As the decades and centuries went by, the mud, sand and rock debris left behind by the melted ice was gradually recolonised by plant life. There was no soil as such, to begin with. But once the lichens, mosses and liverworts became established, organic matter began to be added back into the system and fertility started to build up. Within a few thousand years, soils had formed widely and this great plain, extending out westwards from the Welsh mountains, was green once more. Its inhabitants had taken over what was in essence an ecological blank slate.

The geography of north-west Europe at the Last Glacial Maximum, 25,000 years ago

Collapsing ice sheets and rising seas

When a glaciation comes to an end, a colossal amount of ice is converted into water that flows out into the oceans, so that global sea levels inevitably rise. But there are local factors to take into account, too. If you cover a large enough area of land with a kilometre-thick ice cap, that's a substantial amount of extra weight pushing downwards. Applied over thousands of years, that extra weight is enough to slowly push down the land surface, over a wide area, by as much as a few tens of metres in severe cases. When the ice melts away, all that overhead pressure is relieved and the land surface bounces back — slowly, again over thousands of years.

That bounce-back of the land surface can also determine how much total sea level rise can affect any one area. It's certainly a factor in North Wales, although the effect was at its greatest in Scotland, where there was a much bigger and therefore heavier ice cap. Parts of the world that missed out on the glaciation, and so remained unaffected by the phenomenon, bore the full brunt of the post-glacial sea level change which, on a global scale, is considered to have been some 125 metres. That's nearly 30 metres higher than the tower of Big Ben in London.

How quickly did the change in sea levels take place? Well, such things do not tend to go from A to B in a straight line at a constant speed; instead, changes occur in fits and starts, fast and slow. Up until about 17,000 years ago, little had really changed. This lack of progress was a global, not a local factor. Although the Welsh ice caps had already beaten a retreat, they were tiny in the overall scheme of things. As such, when they melted, their effect on global sea levels was hardly noticeable.

In stark contrast, the gigantic ice caps of the Planet, such as the Laurentide ice sheet, covering much of North America and Canada, were far more stubborn to melt: there was a lot of inertia to overcome. Inertia refers to something that has a resistance to change from its current state, meaning that it's hard to get an ice sheet to start to collapse, but once it does start, that collapse becomes the 'new normal' and it's equally hard to stop.

Sea level change since the Last Glacial Maximum - 25,000 years ago

The dark blue curve represents sea level, as it rose through time. The left-hand map shows Wales and the lands surrounding it at the height of Meltwater Pulse 1A. Although the latter was a dramatic inundation, its effects did not reach Cardigan Bay, which remained land until a few thousand years later.

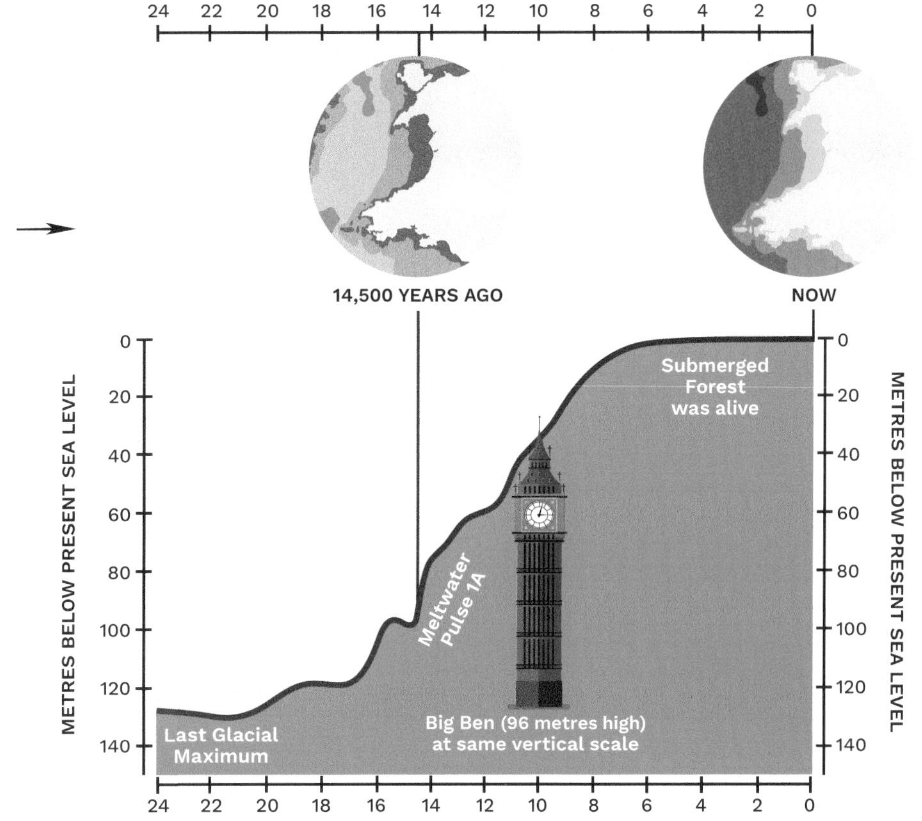

Ice sheet collapse deserves a little explanation. Although it's not something that happens in moments, like when demolition experts bring down a condemned tower-block, in geological terms it is nevertheless a rapid process.

In a warming climate, several combined factors get to work against an ice sheet. A good example is the enhanced seasonal melting of snow and ice, through sunshine, warmth and rain, so that the surface of an ice sheet becomes dotted with pools of meltwater. The pools can be lake-sized affairs in particularly intense melt years and they may become linked together by fast-flowing rivers, carving channels into the surface of the ice.

Water always finds its way downwards through gravity, in the case of an ice sheet via any crevasse it may encounter – and there tend to be plenty of crevasses available. Down the water thunders, deep into the innards of the ice sheet, in turn causing more melting — it can leave parts of the inside of an ice sheet riddled with holes, like a Swiss cheese.

When the water gets down to the base of the ice sheet, it takes up the role of a lubricant, reducing the friction between the ice and the underlying bedrock. That lubrication can speed up the downhill, seaward flow of the ice. Ice sheet collapse involves a combination of warmth, gravity and mechanics that, once well in charge of things, literally drains the ice out from the interior of a land-mass and, in due course, into the ocean.

Sea ice is formed by the sea's surface freezing over. When it is melted, there's hardly any effect on global sea levels — the ice is in the sea already. But when a thick, land-based ice sheet collapses, that's new water being added into the oceans and sea level rise will inevitably occur. The amount and rate of sea level change depends on the size of the collapsing ice sheet and how quickly its demise is brought about.

People always notice the first really spring-like day after the depths of winter and in the case of the transition into interglacial conditions, a similar milestone was reached, some 14,700 years ago. At that point, global sea levels were still 100 metres lower than those of the present day, so that all of Cardigan Bay was still vegetated land. Ynyslas would have been similar in appearance — a wooded lowland, studded with scattered rocky hills, sloping down towards the Dyfi — in other words, nothing like today's familiar sight.

That arrival of much warmer conditions set off a major spate of ice sheet collapse, accompanied by rapid sea level rise. Known to climate scientists as Meltwater Pulse 1A, this event was certainly dramatic in its nature. The oceans rose, possibly by as much as four metres per century, over a period lasting for several hundred years. Despite such drama, though, the green plains of Cardigan Bay remained undisturbed by the waves – for the time being.

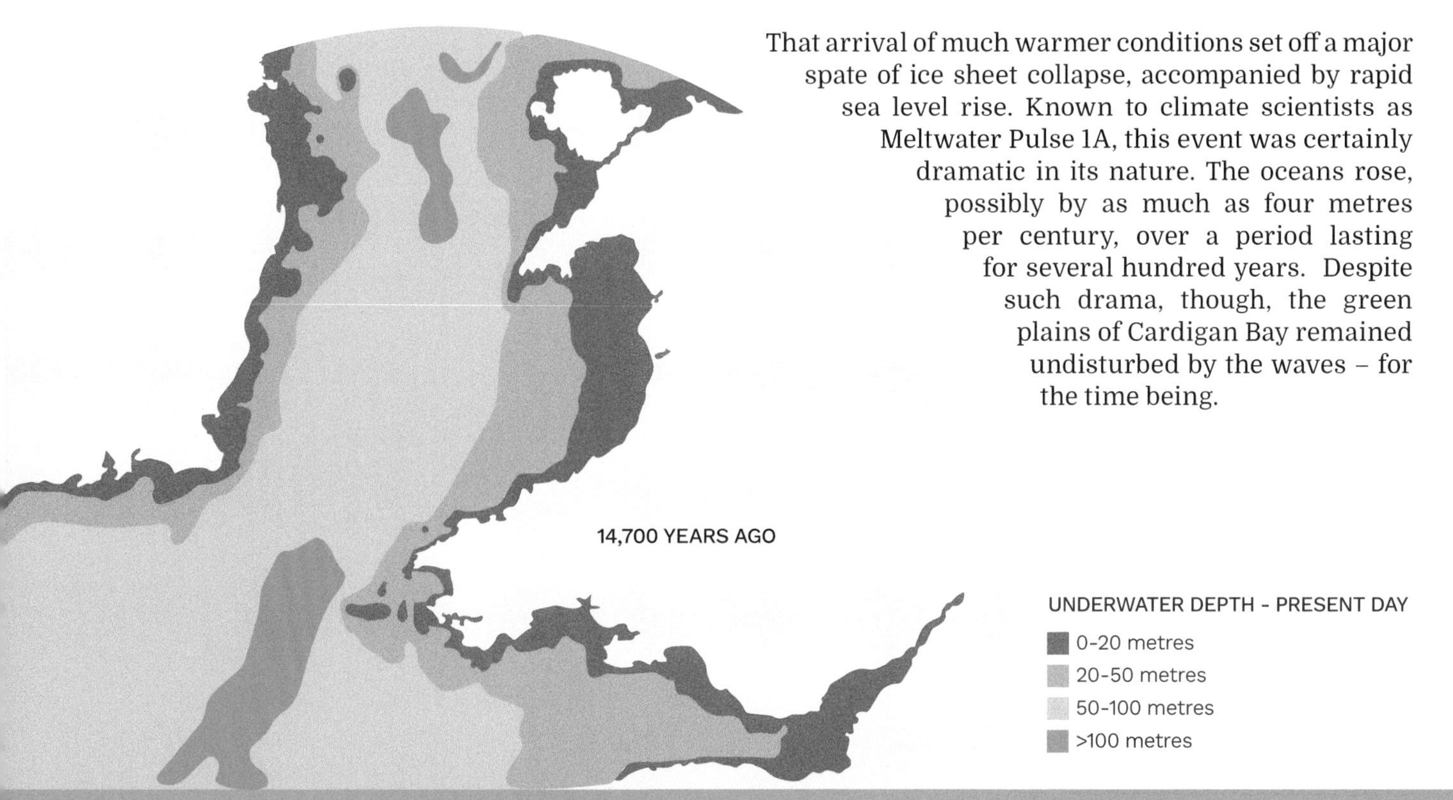

THE MAKING OF **YNYSLAS**

Part two: 11,784 years ago to today - The drowning of Cardigan Bay

Holocene Period

By the start of the Holocene, a division of geological time beginning 11,784 years ago, global sea levels were still some 55 metres lower than those of the present day. In the steadily warming climate that held sway from the early Holocene onwards, the seas rose progressively at an average rate of just over a metre per century, until about 7,000 years ago, when the advance slowed down to a crawl. It was during this critical 5,000 year period that most of Cardigan Bay was drowned.

As the sea flooded in, the deep river-channels cut across the great western plain were progressively silted-up. The waves reworked the sands and gravels previously deposited by rivers and glaciers. Freshwater rivers became estuaries and wooded flood plains became saltmarsh as, century-by-century, the high-water mark edged, mile-upon-mile, towards the modern-day coastline. Marking the open coast would have been a sandy beach backed by a shingle ridge — just like any other exposed shoreline today, but slowly and steadily on the move.

The shingle would have come from those reworked glacial deposits. Only as the coastline approached its present position did the waves start to get to work on a good supply of solid rock — the grey mudstones and sandstones of early Silurian age (438 million years old) that form the sea cliffs we see today between Borth and Aberystwyth. Further out in Cardigan Bay, much younger rocks are present beneath the seabed, as originally predicted by Professor O.T. Jones, a geologist at Aberystwyth University, in 1956. The prediction was subsequently confirmed beyond doubt by drilling a nearly two kilometre-deep borehole at Mochras Point, near Harlech, in the late 1960s. The hole bottomed out in rocks of upper Triassic age, some 225 million years old. Mochras was chosen as the site for the drilling because seismic surveys had indicated that it lay on the seaward side of a massive geological fault, that separated the old rocks of the mainland from the relatively young strata beneath most of the bay. Following that initial discovery, the distribution of these younger rocks was mapped by offshore drilling, so that we now have a good understanding of what lies beneath those reworked glacial sediments.

Some of the offshore boreholes also provided hard evidence of vegetated land out in Cardigan Bay. One was drilled in 1971, at a site 15 kilometres north-west of Ynyslas, in about 20 metres of water. The drill went through 19.5 metres of clays and sands, these being reworked post-glacial sediments, before encountering a metre of brown peat full of wood fragments. The peat is of unknown age but it rested on 22.5 metres of boulder-clay — the stuff dumped by the glaciers as they melted. Beneath the boulder-clay, the hole went into limestones of Middle Jurassic age — about 167 million years old. Another peat-bed was encountered in a borehole drilled in 1972, off Aberaeron, about 35 kilometres south-west of Ynyslas. Sea depth at this site was 18 metres and the peat was encountered two and a half metres beneath the seabed. Radiocarbon dating showed this peat to be some 8,740 years old. Pollen analysis of the peat and the younger, overlying sediments indicated that the ecosystem had changed, over several hundred years, from open, mixed and boggy woodland to saltmarsh.

Relics of a recent past: the Submerged Forest, Borth.

The Submerged Forest

Ancient, albeit somewhat younger peat-beds can also be found along Cardigan Bay's beaches, most famously at Borth's Submerged Forest. Every winter, after storms have scoured away a lot of the sand, great expanses of peat are revealed at low tide. There are countless stumps of long-dead trees, including alder, birch, oak and pine. The bones of an aurochs – a formidable kind of wild ox – were discovered embedded in the peat many years ago and more recent finds have included a magnificent set of red deer antlers.

When the trees were alive, over a 1,300 year period between 6,000 and 4,700 years ago, sea levels were still a little lower than they are now, so the open coast was situated about a kilometre out to the west. During the three millennia or so of geological time recorded in these beds of peat, the coastline advanced slowly but progressively landward. The trees eventually became waterlogged and died.

Detailed investigations of the Submerged Forest, involving the extraction of pollen samples from the peats and clays, have given a good picture of the changes that befell this environment through time. The first thing that happened was the development of a fen-like habitat over once tidal mudflats, with the development of extensive reed-beds. Marshy alder woods followed, and in time the alder trees were joined by birch and pine, with oaks in places at the southern end towards Borth Head, before the area became an acidic sphagnum bog, just like Borth Bog today.

Pinewood in the Submerged Forest, sand-blasted clean by storm waves. November 2009.

The peat-beds at Borth span an important period of human activity in the district. Burnt stone scatters, consisting of numerous sharp, angular fragments of bleached-looking rock, can sometimes be seen in the peat. These fragments were produced by heating-up large pebbles in fires and then immersing them in cold water: it was the sudden cooling that shattered them.

Why people once heated-up rocks in this manner is not clear. One hypothesis is that it was a means of heating-up the water, whilst some other archaeologists have connected the practise with sweat lodges, as in the Native American tradition. Whatever the cause, burnt stone scatters are a reliable indicator of human activity.

A chance discovery, made in 2012, in the surface of one of the peat-beds, was even more remarkable: footprints. There are circular depressions, several centimetres deep and made by the hooves of cattle (or perhaps aurochs), plus the sharp, cloven hoofprints of deer, sheep and goats and, incredibly, human footprints. It appears that the adults were foot-clad, but one print was found that belonged to a child around four years old, showing the impressions of bare toes.

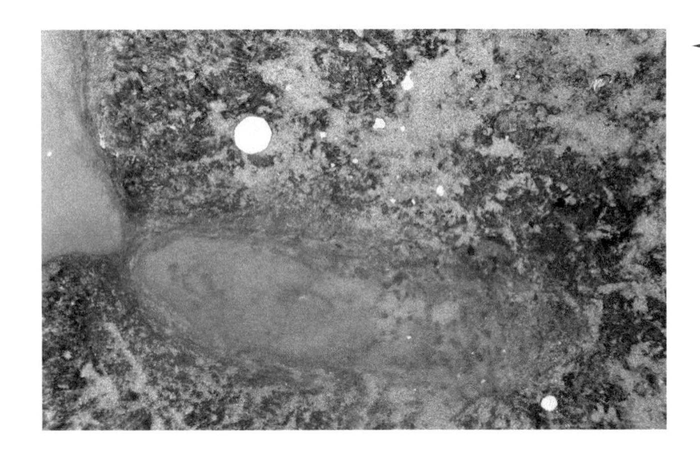

A human footprint at the Submerged Forest, Borth, March 2012.

Photographed in February 2014, after severe winter storms stripped the beach of much of its sand cover. 50p for scale, top centre.

Ancient hoofprints, both rounded and cloven, in the peat at the Submerged Forest, Borth.

Today, even if you stamp hard with heavy boots, you will barely scuff the peat surface. We normally think of peat as soft and squidgy when wet, so why, when it was once easy to make footprints in these beds at Borth, is it impossible now? As with all physical features of Earth's surface, the reason is to be found in geological processes, in this case the landward advance of the coastline. To appreciate how this happened, one first needs to consider how the system formed and how it is maintained.

Even the tracks of a heavy excavator make little impression on the compacted peat of the Submerged Forest.

Storm beaches and sand dunes

If you visit the Submerged Forest at low tide, look to the north, towards Aberdyfi. The view takes in the expansive, golden ripple-marked intertidal sands, towards the long, steep shingle-bank forming the beach-head, with the sand dunes standing beyond like a mini mountain range — a classic storm beach and dune system. Such systems all work in a similar fashion, so their basic requirements are essentially the same. They must have a continuous supply of fine sand, together with a prevailing onshore wind that consistently transports the sand inland to be added to the dunes. Sand is not a problem, since reworking of the glacial debris by the waves has provided ample resources. One only needs to look out over the Estuary mouth at low tide, with its maze of large and constantly shifting sand-banks, to appreciate that.

From the top of one of the bigger dunes, looking down onto the beach at low tide and out to the shifting sandbars and snaking channels of the outer Estuary of the Dyfi. →

Sand dunes north of Aberdyfi, where there is less of a protective storm beach, were hammered by the great storms of early 2014. →

Ynyslas must also have a supply of coarser rock-debris to maintain the storm beach — a suit of armour that protects the dunes from being destroyed by the huge waves that batter the coast during winter storms. The maintenance system works thus: frequent cliff-falls to the south of Borth supply a feed of raw materials in the form of rock-debris. When the weather is rough and the waves have the most energy, they pound the debris together, gradually breaking it up into more manageable and transportable chunks. Since both the prevailing wind and the flood-tide come along the coast from the south-west, the rocks are transported in a northerly direction, right along the beach. Such erosion has been ongoing for a few thousand years, with the slowly retreating cliff bases having been progressively worn down into extensive, flat-lying intertidal reef systems.

An active rockfall under the headland at Borth - one of the stone-supply sources for the storm beach.

As waves roll and bash the rocks around, over time they become the rounded pebbles and cobbles that make up the shingle bank. Today's storm beach is dominated by grey pebbles from the cliffs, but if you look more closely, you will also find a variety of other rock types. Milk-white quartz is common and much of it is local, since the Silurian rocks contain a lot of quartz veins. From further afield come all sorts of greenish-grey, crystalline volcanic rocks from Snowdonia, eroded from the mountains and dumped on the floor of Cardigan Bay by those ice age glaciers. Pebbles of more exotic origin have been reworked from the rubble left behind by the retreating Irish Sea ice.

During the recent sea defence construction at Borth, many lorryloads of coarse shingle were brought in to add bulk to the storm beach there. These came from gravel pits on the Llyn Peninsula, parts of which were over-ridden by the Irish Sea ice and left covered in glacial detritus: as a consequence, there is more of an Irish Sea component to the pebbles now than there used to be. Because the Irish Sea ice flowed all the way down the western side of Britain, the debris it left behind contains a vast diversity of rock types. Pale granite from Scotland and bright red jasper from ancient rocks on the Llyn Peninsula are less common, but they are particularly easy to spot, especially when the pebbles are still wet on an ebbing tide.

Ignimbrite, a common middle Ordovician volcanic rock and formed by catastrophic eruptions like the one of Mt St Helens in 1980.

Lower Silurian siltstone with abundant white quartz veins.

Tuff - a rock made up of volcanic ash, ejected in explosive eruptions in the Ordovician Period, some 460 million years ago.

Dark grey basalt lava, in which gas bubbles were mineralised and are filled with calcite (white).

Breccia (pronounced brechier) - fragments of Silurian sedimentary rock (dark) in quartz (white).

Paler grey lower Silurian siltstone.

Lower Silurian siltstone with pale brown dolomite and whitish quartz veins.

Dark grey lower Silurian mudstone, some 440 million years old.

Porphyry (pronounced por-fer-ee) - an igneous rock consisting of larger, visible crystals within a finer-grained groundmass.

Fine-grained blood-red jasper, largely consisting of silica and red hematite (iron oxide).

That's how the storm beach works: now to the dunes. When the tide is out and the weather is dry but with strong onshore winds, the surface of the beach dries out and sand is blown inland. As well as adding more sand to the existing dunes, new dunes may start to form wherever there is even slight shelter from the wind; as the sand accumulates it creates still more shelter, so dune formation is a self-perpetuating process. A newly formed dune has a lot to compete with, though. Very high tides, caused by rare storm-surges, may obliterate a young dune. But if a new dune can survive for long enough, it will begin to be colonised by pioneering plants. These are seriously tough cookies, with waxy coatings on leaves and stems to stop them drying out, and an unusually high tolerance to the saltiness and extremes of wet and dry, heat and cold that come with this hostile territory, exposed to all weathers. Their roots begin to knit the sand together, reducing erosion and encouraging further accumulation. Such protection is vital, since if a storm-force wind gets into a dune, the loose sand can be blasted apart in a matter of hours.

When the wind gets into sand dunes, it can rip them apart.
Photograph taken during storm-force 10 winds, November 2013.

Once a young dune is stabilised in this way and begins to grow, it forms a perfect habitat for the plant that epitomises sand dune scenery, Marram grass. Marram is so vigorous that, once established, it can keep up with rapid accumulations of sand – if it gets buried, it just grows a bit more and pops back up again. It has deep roots that can get down to moisture in the driest of summers, so for a time it will out-compete pretty much everything else. But eventually, an old, totally stable dune begins to develop a thin soil and a variety of other plants will find a home. At Ynyslas, many of these plants are those with a liking for a bit of lime, since the sand is a mixture of tiny rock and mineral grains and countless fragments of seashells, made of calcium carbonate. Some of these plants like the well-drained dunes themselves, while others, such as most of the orchids for which Ynyslas is famous, prefer the flat-lying slacks in between the dunes, where the ground surface is much closer to the water table, so that conditions tend to be on the damp side.

This spectacular orchid, common in the relatively moist dune-slacks at Ynyslas, is known as the Leopard, from the distinctive dark markings on its leaves. A variety of the Southern March Orchid, it is one of several orchid species that can be found in bloom here in late spring and early summer. →

The shingle storm beach is at least two metres high and many metres in width, with the dunes forming a tract 200 to 500 metres wide and, in their most stable interior parts, up to 10 metres in height. When the Submerged Forest was a living place, the open coastline was located about a kilometre offshore, yet it now lies just landward of the remains of the trees. Over the past 7,000 years, following the slowdown in sea level rise, it has very slowly crept to its present position. It logically follows that, during that advance, there must have been a considerable amount of time during which the storm beach and, very probably, some form of dune system, were situated smack-bang on top of what is now the visible part of the Submerged Forest. With several metres of sediment sat on top of it, the peat-beds had to bear many tonnes per square metre of overhead pressure. The peat became compacted, and that's why your boots now leave no mark.

Such compaction happens to all progressively-buried sediments. It is the initial stage of the geological transformation that, given enough time and ever-deeper burial, eventually turns soft sediments into solid, hard sedimentary rocks like sandstone – and peat into coal. But thousands of years ago, when those footprints were made and the coastline was still some distance out to the west, no such compaction had occurred; what is now the Submerged Forest would have been a marshy woodland, where cattle wallowed and sheep watered – and where people either dwelt or visited.

Here's an interesting geological conundrum. After a big storm, when the Submerged Forest is really well-exposed, the peat-bed can be seen to be underlain by grey clay. The clay contains the remains of bivalves and other marine organisms, indicative of an intertidal estuary mudflat habitat, a bit like some of the squelchier parts of the Dyfi Estuary today. This apparently back-to-front sequence – estuarine conditions followed by the development of a peaty forest – seems a bit of a contradiction when the overall theme is one of progressively rising sea levels. But remember that some 7,000 years ago, a few centuries before the Forest became established, the rate of global sea level rise had dropped off sharply.

After major storms, when more sand is scoured away, tidal-flat clays can be seen to be underlying the peat.

Geological problems like this are good fun and various explanations have been offered over the years to account for the pattern. Did sea levels or tides oscillate a bit? Not so far as we know – but that doesn't necessarily mean they didn't. Did the land rise? It seems unlikely – only in northernmost Wales do we see good evidence for significant post-glacial land uplift. How about the shingle spit? This may well have played a role. The advancing coast was situated a bit further out at the time, but once it had reached the outcrops of grey Silurian mudstones, today represented by Borth Head, a cliff-line would have been formed and then gradually eroded back inland towards its present position. That development would have greatly increased the rock supply to the spit, which would have grown in height and, perhaps, length, significantly improving protection from the waves. As a consequence, the sharp slowdown in sea level rise, the continued transport of sediment down the rivers and the improved shelter afforded by the expanding barrier of the storm beach, may well have worked in combination. In this scenario, large parts of the intertidal mud-flats, inland of the shingle-spit, would have silted up completely, so that for many centuries, the habitat once again became non-marine. Clearly, this was only a temporary state of affairs, since as you will note when visiting the Submerged Forest, the sea won, in the end.

Although on a daily basis, Ynyslas looks just the same as it did the day before, changes are again afoot. The vital northwards transport of pebbles from the cliffs at Borth, feeding the storm beach, has slowed down. Along some sections of the storm beach at Ynyslas, especially to the south of the dunes, the picture now is one of active erosion, something likely to increase in due course thanks to modern sea level rise. It is a difficult balance to strike: on the one hand, there is the very real need to try and protect Borth from flooding caused by storm-waves overtopping the seawall, as happened in winter 2013-14. On the other, it must be remembered that the storm beach is the protective armour of this strip of coast, so it needs looking after. Man versus sea: it's an old theme and in most cases the winner is ultimately the sea.

The storms of winter 2013-14 were some of the worst to hit Cardigan Bay in living memory.

The conversion of that great western plain into Cardigan Bay, as we know it today, is one such case and people may well have witnessed the process. Critically, the Mesolithic, or Middle Stone Age, began 11,600 years ago, just into the Holocene, and lasted for some 5,000 years. There are a number of sites around Wales where artefacts from this time have been recorded. People were certainly around. Any Mesolithic folk present in western Wales would have noticed, as the decades and generations passed, the slow but steady inundation of Cardigan Bay. As the coastline advanced, gradually but steadily towards the mountains, those expansive lowlands, that had endured for thousands of years, shrank and shrank.

We know, from the remains preserved in the Submerged Forest, that the great western plain would have featured open woodland of oak, pine, birch, alder, hazel and other species, inhabited by grazing, browsing and rooting animals like wild boar, aurochs, red and roe deer. By the start of the Mesolithic, there was no longer a megafauna: it had long since disappeared, hunted into oblivion. The largest predators that remained were wolf and brown bear. European peoples of this time seem to have been part-nomadic, as the seasons permitted, and part-occupiers, so that suitable spots were seasonally or even permanently inhabited. The wooded, food-rich lowlands that are now Cardigan Bay would have provided excellent opportunities for such settlement, so long as one sensibly avoided areas that were too low-lying or too close to rivers and therefore prone to flooding.

It does make one wonder: is there ever smoke without fire? There is a very interesting legend about Cardigan Bay: the tale of Cantre'r Gwaelod.

← This image shows the road from Tywyn to Tonfanau, just up the coast from Ynyslas, after the storm of January 3rd, 2014. The Cambrian Coast railway runs along the top of the storm beach here for about a kilometre. In just a couple of hours, the huge waves swept aside the track-ballast, along with boulders and a stout concrete and steel fence, completely burying the road.

THE MAKING OF **YNYSLAS**

Part three: Global sea level rise - the origin of the Flood Legends?

Cantre'r Gwaelod, translating into English as the Lowland Hundred, is the legendary, now sunken kingdom that is said to have once occupied a tract of fertile land, in what today is Cardigan Bay. The legend speaks of a main settlement called Caer Wyddno, seat of the ruler Gwyddno Garanhir. In some versions of the tale, the ridge-like glacial moraines that make up the Sarnau of Cardigan Bay have become the causeways heading out to the settlement. The land is described as having been defended from the sea by a great earthwork or dyke, complete with sluice gates, that could be opened at low water, but which needed to be shut fast when the tide came back in.

Apparently, the earliest written version of the legend is the poem, Boddi Maes Gwyddno, in the Black Book of Carmarthen, dating back to the year 1250 or earlier. In this version, the land, called Maes Gwyddno, was visited by Seithennin, who was a neighbouring king. He became attracted to the "well cup-bearer", Mererid. The poem translates thus:

Seithennin, stand forth
And behold the seething ocean:
It has covered Gwyddno's lands.

Cursed be the maiden
Who let it loose after the feast,
The cup-bearer of the mighty sea.

Cursed be the girl
Who let it loose after battle,
The cup-bearer of the desolate ocean.

Mererid's cry from the city's heights
Reaches even God.
After pride comes a long ending.

Mererid's cry from the city's heights today,
Implores God.
After pride comes remorse.

Mererid's cry overcomes me tonight,
And I cannot prosper.
After pride comes a fall.

Mererid's cry from strong wines;
Bountiful God has made this.
After excess comes poverty.

Mererid's cry drives me
From my chamber.
After pride comes devastation.

The grave of high-minded Seithennin,
Between Caer Genedr and the sea:
Such a great leader was he.

- **The Black Book of Carmarthen** (translation - https://cy.wikipedia.org)

The poem implies that Mererid was responsible for the sluice gates and neglected such matters during the feasting and merriment. In another popular version, though, Seithennin is the gate-keeper and on a dark and stormy night, after a few drinks too many, he neglected the gates. As the tide rose, the sea swept in, flooding the land.

There always seems to be a moral involved somewhere. It is said that the modern version of the legend dates from the 17th Century, but was later popularly adapted by the Temperance Movement, as an example, according to them, of the dire consequences of non-temperance.

But is that poem in the Black Book just a written and fragmentary interpretation of something much, much older, originating via the oral tradition? Is the tale of Cantre'r Gwaelod a variously-embellished account of the slow but relentless inundation of a great fertile plain, over several millennia?

Traditional, independently-evolved ancient Flood Legends exist in cultures the world over, something that is suggestive of a global event that took place in the distant past. A large, global and at times rapid rise in sea level certainly happened over the many millennia following the Last Glacial Maximum. The only unanswered question is: did people talk about it? We would. Can you imagine? We would be posting images and videos of coastal flooding all over social media.

Although sea level rise of a metre per century would have been imperceptible on a day-to-day basis, over an intergenerational scale it would have been blatantly obvious. The more bountiful the land had been, the greater the loss, if one reasonably equates fertile lowland ground with food and shelter.

Just how far back does the oral tradition go? It must surely have originated as soon as humans developed languages with which to communicate and, vitally, co-operate. You can almost imagine a bunch of Mesolithic folk sat around the fire, an elder telling a young man that in his Great-Grandfather's day, the hunting was twice as good because they had far more fertile land to work with.

Science seeks its answers via hard evidence, often with success, but in some fields of study the picture is incomplete. In the case of Cantre'r Gwaelod, on the one hand, we have an old flood legend, one specific to Cardigan Bay, but which is echoed in independently-evolved folklore all around the world, embellished no doubt through time, but unavoidably there. On the other hand, we have the hard evidence for Cardigan Bay having been land into the Mesolithic era and having been subsequently drowned.

A truly mountainous sea during storm-force 10 winds, Ynyslas, November 2013. Shut those sluice gates!

Will the two threads ever be joined? It's something to think about when contemplating the Submerged Forest, especially at sunset on a flooding tide, the incoming sea swirling around silhouetted tree stumps protruding dark from the water, until they vanish from sight as night gathers swiftly over the waves. This is the sea's domain now, regardless of whose abode it was, all those thousands of years ago.

Could we be risking a Cantre'r Gwaelod 2.0, through our reckless burning of the fossil fuels? The laws of physics say yes. Temperatures are rising. Sea level rise, currently at about three and a half millimetres a year, is accelerating. The world's great ice caps contain enough water to raise sea levels by 66 metres or so, if they melted completely. So once again we need to look at the past when considering the future. The last time atmospheric carbon dioxide levels were at 400 ppm plus, as they are now, was some 3.5 million years ago, during a stage of geological time known as the Pliocene.

In the Pliocene, 5.33 to 2.58 million years ago, carbon dioxide levels were instead falling away from even higher values. Following tens of millions of years of Hothouse warmth, the climate was heading towards the Icehouse of the Quaternary. Before conditions had cooled too much, though, in the Mid-Pliocene, forests still flourished inside the Arctic Circle, where all that we see now is stunted tundra, and as recent research has shown, there were areas of woodland on Antarctica. Global sea levels were significantly higher. That's a snapshot of a 400 ppm CO_2 Earth when it's cooling. In our case, having already created a Mid-Pliocene atmosphere, it seems we're not satisfied with that. Should our carbon dioxide emissions continue at this rate, over the coming decades, we will steadily drive our atmosphere towards one last known on Earth many millions of years ago, when the Polar ice caps were much smaller.

But unlike past Icehouse-Hothouse climate transitions, this one is happening very quickly indeed. On a geological time-scale, it is almost instantaneous, thanks to Man's industrial, domestic and recreational activities of just a few centuries. Man-made climate change is entirely our problem and we need to solve it, now. What will people be writing about us in a thousand years' time? What sort of description will we have earned for ourselves in that account?

That, my friends, is entirely up to us.

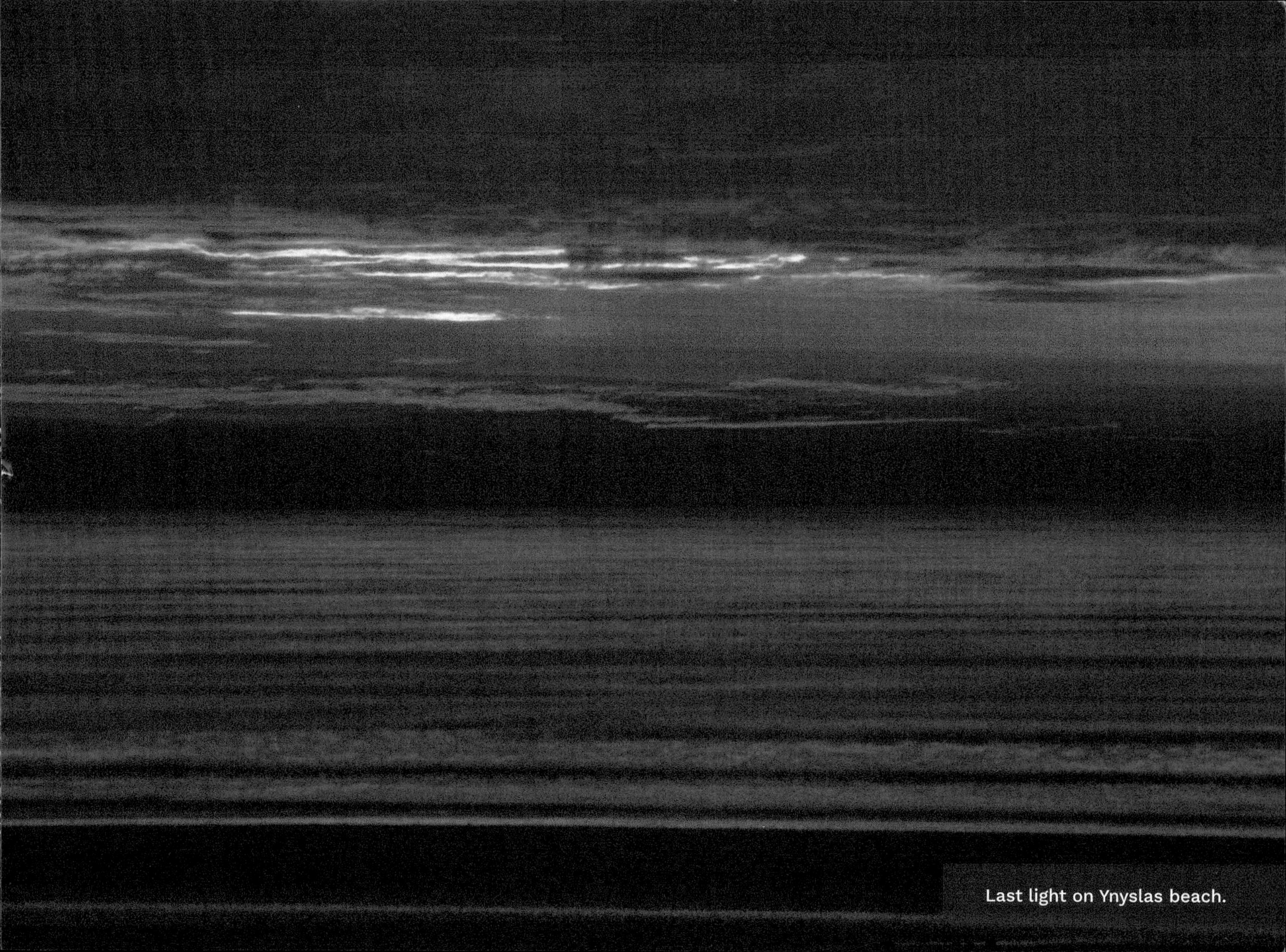

Last light on Ynyslas beach.

Cantre'r Gwaelod

O, Seithennin,
You, who are forever blamed in legend
As they tell of that one too many
Quite fond of the stuff you were by all accounts
And as you worked your way
Towards the wrong side of another night
In it came and out you went

You won't stop the sea

O, Seithennin,
You won't stop the sea
You could not have stopped the sea
As the continents of the North shed their ice
And you fought the change
On your fragile plain
As on it came

You won't stop the sea

O, Seithennin,
Were you castigated back then as an ecological monger of doom?
And was ridicule your lot in the face of the coming danger?
Just supposing you had seen into the future
Seen what we have now found
How the sea came on regardless
How Cantre'r Gwaelod was drowned

You won't stop the sea

O, Seithennin,
I have cast from shorelines into your deep
I have sailed over your plain and beyond
On echo-sounder screen I have seen the shape of your hidden land
Where sea-bream now shoal
What lies, now submerged?
Fact and fantasy - have they merged?

You won't stop the sea

O, Seithennin,
You who are forever blamed in legend
Rest, my friend, it was just climate change
An inevitable consequence of the laws of physics
The dynamics of Planet Earth - if you like.
They won't have it now - as they wouldn't have it then
Though as you found out: it wasn't if - but when...

You won't stop the sea

And – yes the sea relentless came in
Legend tells, at your hand
And though the cause lay beyond
They did not understand
It's high water at Borth
Summer's going as planned
And now the jetski holds sway
Over Garanhir's land

John S Mason

Acknowledgements

This book is long in the making. It began life in 2014 as a detailed analysis of our geological origins here in Wales. Intended as a tribute to my late mother and the great encouragement that was given, by her, my father and grandparents, to a young mind curious about all things wild and especially fossilised, it became something of a journey of discovery, through aspects of the Earth sciences that, as a mineralogist by speciality, I had not entirely appreciated. What was I to do with this now-accumulated mountain of information? Rather than seek to publish it all in a single weighty volume, I realised it would reach a wider readership if it could be sliced up into bite-sized portions, with the focus each time on a specific, well-loved place in the Welsh Landscape. The next slice is coming soon, but for now, I hope that you enjoy this part of the story. If it can ignite the spark of curiosity in one young mind, then I will be happy.

For their constructive reviews of earlier drafts of the text and their help with selection of graphics, major gratitude to Doug Bostrom, David Kirtley, John Garrett and Dan Bailey, from the Skeptical Science team (https://skepticalscience.com).

Thanks to my fantastic designer and illustrator, Sara Holloway, for her diligence and timely work. Thanks also to my very patient editor, Fiona Mason, for her guidance and indeed inspiration into getting this project off the ground and keeping it pointing in the right direction!

John S Mason, MPhil, is a geologist by training with a long-held love of wildlife and a deep interest in weather and climate. He has been walking, sea-fishing, photographing landscapes, wildlife and weather, day and night around Ynyslas for over 35 years. On climate change, he writes for the award-winning website, Skeptical Science, with a particular focus upon research into climates of the past. He is also the author of the successful *Shore Fishing - a Guide to Cardigan Bay* (Coch y Bonddu Books, Machynlleth, 2013), *Introducing Mineralogy* (Dunedin Academic Press, 2016) and numerous peer-reviewed contributions to British Mineralogy, including co-authorship of the major volume, *Mineralization in England and Wales* (published by the Joint Nature Conservation Committee).

www.geologywales.co.uk